DAMIANI

DAMIANI©2005

DAMIANI EDITORE
Via Zanardi, 376
Tel. +39.051.6350805
Fax +39.051.6347188
40131 Bologna - Italy
www.damianieditore.it
info@damianieditore.it

Printed in Italy
Grafiche Damiani s.r.l.
www.grafichedamiani.it

Printed on
Magno Satin 170gr.
distribuited by

antalis EM
antalis.com

Author notes
This book is unauthorized and
the author has received no help
in any way from Omega or any of its
associated companies or any serving employee.

OMEGA is registered trade mark of
OMEGA S.A. SWATCH GROUP®
The following marks are property to OMEGA S.A.:
Omega Art, Co-Axial, Constellation,
De Ville, Dynamic, Seamaster, Speedmaster and X-33
All trade marks are used for information and
identification purposes only and
no endorsement is implied.

OMEGA
WATCHES

Photografed and written
by
JOHN GOLDBERGER

Introduction
by
GIAMPIERO NEGRETTI

DAMIANI

INTRODUCTION

A book about a make of watches may be conceived in a variety of ways. The simplest is that of gathering the contents from official sources, museums and historical archives, thus retracing – with a broader perspective at times – the path already defined by others. Instead, a lengthier, more difficult and more demanding alternative is that of performing what is usually referred to as field research: seeking out unreleased or little known models, or those only identified to date through the images included on pricelists or advertising material from the period. This latter is a sort of hunt requiring patience, intuition, knowledge and competence. And this is the approach opted for by John Goldberger, who for many years has been attending auctions, enthusiastically sifting through antiques markets, conventions and visiting specialist shops in all four corners of the world. I have known John Goldberger for a long time now – almost twenty years – and I should firstly say that behind his "nome de plume" lies a great collector, a person with a deep-rooted love of watches. A great expert and one who understands the two 'souls' of a timepiece: the exterior one (the appearance given by design) and the more concealed one (the mechanics).

Differently to many other enthusiasts, John does not simply collect period watches for the speculative pleasure – something that was certainly not uncommon in the years when he began frequenting the 'watch world' – but rather due to his wish to gather together exceptional pieces, whether they be enhanced by a renowned signature, by the item's individual history, by its unusual style, or perhaps by outstanding technical features. This passion in fact drives him to patiently seek out excellently preserved examples or even never used or practically new period pieces. And this is the background to the birth of the book Omega Watches focusing on this make. The publication enables us to admire 250 interesting and attractive watches, including many rare examples – ones not even to be found in the Omega museum – most of which belong to John Goldberger's private collection. Mr. Goldberger's hand is not only responsible for the technical descriptions in the book, but also for the images: photography is in fact another of his great passions and has seen him as the editor and photographer behind a stimulating book of travel images.

Established in 1848 in La-Chaux-de-Fonds by Louis Brandt and later moved to Bienne (its current premises) in 1880, Omega became the company it is today only at the beginning of the last century. Indeed, prior to this the Louis Brandt products were known through the various make names that over the years were displayed on the watch faces – Jura, Patria, Helvetia and Labrador were among these.

The name we know today was given by a of 19''' movement produced in 1894; it became so famous for its precision and reliability that the Omega name was included in the company title as of 1903, ousting all the other makes and becoming "the firm". And the book in fact sets off with the early 1900s, presenting a series of pocket-watches, sweeping through a variety of pieces, to finish on some 1970s quartz models such as the Megaquartz 2400 Marine Chronometer – the most precise wristwatch in the world.

This time-span encompasses practically all the most prominent models produced by the Bienne firm: they range from its first military watches to its split second chronographs with measurement for sports competitions, from its square models to the oval ones, from the very rare two-button stopwatches (one button on the side and the other between the lugs) with the same functions as those – just as rare – pieces with the seconds shown both in the dial centre and bottom, and from the watch General Franco gave the President of Mexico, right up to the extensive range of Speedmasters (with extremely rare models preceding Man's first landing on the moon), Seamasters, sophisticated Constellations and the various chronometer with observatory bulletin reports. There are full-page colour illustrations of all the timepieces, with their movements set out alongside and accompanied by a technical description and reference number: everything a true enthusiast can ask for. Thus he or she may observe the most minute case and dial details while also looking at the (enlarged) image of the calibre used and the relative specifications. The graphics chosen are simple and elegant, emphasising visibility (one watch per page), whilst the images and printing – by the Bologna-based Damiani publishing house – are accurate and excellent quality. It is of course a book focusing on the creations produced by one of the most renowned makes in the world, but it is also a book dedicated to all those who succumb to charm and appreciate a beautiful timepiece, whatever name may be on the face. The watch is an instrument that will probably be seen by future generations as something of a dinosaur undergoing extinction. However, as Alfonso Longo, a Lombard exponent of the Enlightenment, wrote in the mid 1700s: "without counting the clocks mounted in earrings or rings, and those marking not only the hours but also the minutes and giving a delightful concert every hour, and those showing the days of the month, the holidays and the planet's paths, and with a singular mastery, both in terms of precision and external craftwork; and furthermore I say, without counting the most celebrated clocks of Europe, to my vision the discovery of the alarm clock or the watch deserves the greatest applause of all inventions".

Giampiero Negretti

Dedicated
to my parents

INDEX

Golden silver pocket watch
Four- body hinged case ø 55 mm.
Enamel dial
Arabic hours and minutes track
Cal. 19''' LC quality C

Stainless steel pocket watch
Four-body hinged case ø 48 mm.
Enamel dial
Black enamels Arabic numerals
Cal. 38 CL T1

Two colours 18 kt gold pocket watch
Four-body "demi-bassine" case ø 47 mm.
Silvered dial
Arabic numerals
Cal. 38 M.S

Platinum openface dress watch
Three-body case ø 45 mm.
Silvered dial
Blue Arabic numerals
Cal. 39.5 L

Stainless steel wrist watch
Three-body case with hooded lugs ø 31 mm.
Enamel dial
Radium Arabic numerals with red minutes
Cal. 26.5

Golden silver pocket watch
Four-body hinged case ø 51 mm.
Enamel dial
24 h black Arabic numerals
Cal. 19''' LB

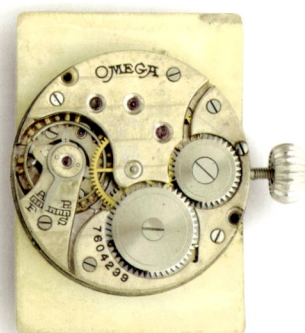

Silver wristwatch
Two-body case with hinged lugs 42x30 mm.
Silvered dial
Luminous baton numerals
Cal. 26.5

Silver wristwatch
Two-body case with hinged lugs 44x25 mm.
Silvered dial
Luminous baton numerals
Cal. 20 F

18 kt gold wristwatch
Tonneau-shaped hinged case 35x24 mm.
Guilloché dial
Painted radial Arabic numerals
Cal. 23.7 S

18 kt gold wristwatch
Two-body hinged case ø 41 mm.
Two-tone gilt dial
Roman numerals
Cal. 26.5 T2 SOB

14 kt gold wristwatch
Tonneau hinged case, 40x31 mm.
Guilloché dial
Engraving black radial Arabic numerals
Cal. 26.5 SOB

Stainless steel aviator wristwatch
Three-body case with screw back,
revolving bezel, ø 41 mm.
Black enamel dial, radium Arabic numerals
Cal. 35.5 S

Stainless steel hour angle aviator wristwatch
Three-body case with 12 hours revolving bezel,
ref. CK2000, ø 44,5 mm.
Black dial with seconds track, radium Arabic numerals
Cal. 44.5

Stainless steel hour angle aviator wristwatch
Three-body case with 12 hours revolving bezel,
ref. CK 2042, ø 41 mm.
Black dial with radium Arabic numerals
Cal. 26.5

Stainless steel dress watch
Three-body case, ø 45 mm.
Two-tone silvered dial
Arabic numerals
Cal. 37.5 T

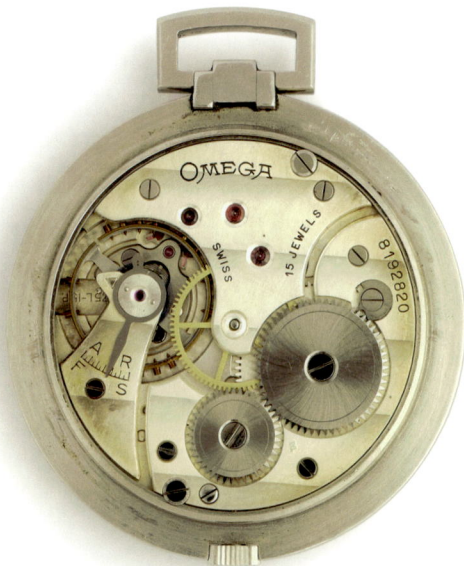

Stainless steel dress watch
Three-body case, crown at 6, ø 46 mm.
Silvered dial with gilt track
Deco arabic numerals
Cal. 37.5-15P

18kt gold dress watch
Elongated case with "champlevè" enamel, 48x42 mm.
Two-tone silvered dial
Arabic numerals
Cal. 23.7

18kt gold dress watch
Engraved tortue shaped case, ø 45 mm.
Glossy silvered dial
Engraved Arabic numerals
Cal. 23.7 L

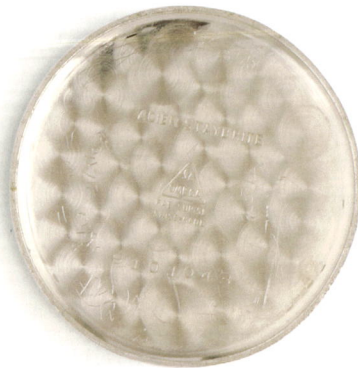

Stainless steel wristwatch
Two-body tonneau case, ø 30 mm..
Enamel dial
Black Arabic numerals
Cal. 26.5

Stainless steel wristwatch
Three-body case, ø 37 mm.
Black dial
Gilt Arabic numerals
Cal. 30

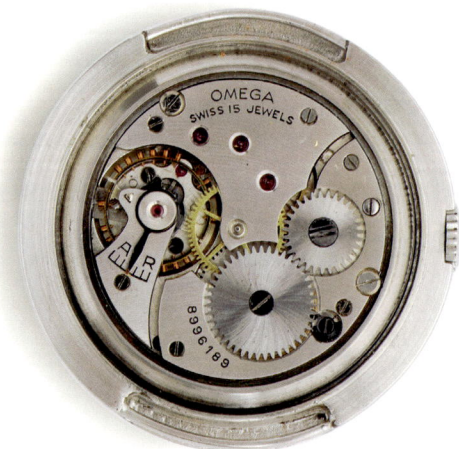

Stainless steel wristwatch
Two-body case, ø 35 mm.
Silvered dial
Arabic numerals
Cal. 26.5

Stainless steel wristwatch
Two-body case, ø 37.5 mm.
Silvered dial
Roman numerals
Cal. 26.5

Stainless steel and chromium-plated wristwatch
Three-body oversized case, ø 44 mm.
Two-tone silvered dial
Arabic numerals
Cal. 37.6 S

Stainless steel dress watch
Three-body case, ø 56 mm.
Two levels black dial
Gold batons with 24 h Arabic numerals
Cal. 37.5 – LT1

Stainless steel wristwatch
Three-body case, ø 37 mm.
Two-tone silvered sector dial
Arabic and baton numerals
Cal. 30

Decò desk clock
Stainless, crhomium-plated case, 225x115 mm.
Silvered dial
Roman numerals
Mov. 8 days Cal. 59.8 D

Stainless steel wristwatch
Three-body case with Omega bracelet, ø 37 mm.
Silvered dial
Radium Arabic numerals with seconds track
Cal. 231

Stainless steel dress watch
Four-body hinged case, ø 46 mm.
Two-tone silvered dial
Arabic numerals
Cal.37.5 L1SP

Brass table chronometer
Two-body case, ø 68 mm.
Silvered dial
Arabic numerals
Cal. 40.6 LT2 17P

Oxydated brass motor clock
Three body case, ø 84 mm.
Black dial
Arabic luminous numerals
Mov. 8 days Cal. 59.8D

Stainless steel wristwatch
Three-body case, ø 37 mm.
"Bull-eye" gilt dial
Arabic numerals
Cal. 30

Stainless steel wristwatch
Three-body case, ø 37 mm.
Silvered dial
Arabic numerals
Cal. 30

Stainless steel wristwatch
Three-body case with teardrop lugs, ø 37 mm.
Black dial
Radium Arabic numerals with seconds track
Cal. 23.4 SC

Stainless steel dress watch
Three-body case, ø 45 mm.
Black sector dial
Silvered Arabic and baton numerals
Cal. 37.5 L 15R

Stainless steel dress watch
Three-body case, ø 48 mm.
Black sector dial
Gilt Arabic and baton numerals
Cal. 37.5 L 15R

Stainless steel dress watch
Three-body case, ø 56 mm.
Two-tone silvered dial
Arabic numerals with exterior second track
Cal. 30 T2 SC

Stainless steel wristwatch
Screwed back case ref. CK2179/3, ø 35 mm.
Two-tone silvered dial
Arabic numerals with exterior second track
Cal. 30 T2 SC

Stainless steel wristwatch
Three-body case, ø 36 mm.
Two-tone silvered dial
Roman numerals
Cal. 30 T2

Stainless steel dress watch
Three-body case, ø 45 mm.
Pink dial with white track
Roman numerals
Cal. 37.5 L 15R

Stainless steel dress watch
Three-body case, ø 45 mm.
Silvered dial
Roman numerals
Cal. 37.5 L 15R

Stainless steel wristwatches group
Original tags with reference and price
Two-tone silvered dials
Cal 30 T2

Stainless steel wristwatch
Three-body case, ø 38 mm.
Silvered dial
Steel applied baton numerals
Cal. 30 T2

18 kt gold wristwatch
Three-body oversized case, ø 44 mm.
Silvered dial
Gold applied baton numerals
Cal. 601 SC

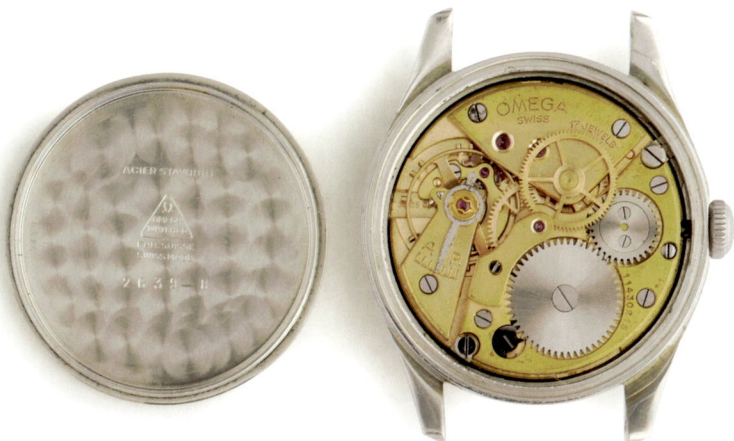

Stainless steel wristwatch
Screwed back case ref. CK2639-8, ø 35 mm.
Silvered dial, sweep center and subsidiary seconds
Steel applied Arabic and baton numerals
Cal. 30 T2 SC-PC

Stainless steel wristwatch "Ranchero"
Two-body waterproof case ref. CK2990, ø 35 mm.
Silvered dial
Arabic numerals and radium batons
Cal. 267

Gold filled and stainless steel wristwatch "Ranchero"
Two-body waterproof case ref. BK2990 ø 35 mm.
Silvered dial
Gold applied Arabic and baton numerals
Cal. 267

Desk clock
Gold plated brass case, 160x155 mm.
Silvered dial
Applied Arabic and baton numerals
Cal. 53.7 SC 6720

14 kt gold wristwatch "Cosmic"
Three-body case ref. OJ2473, ø 35 mm.
Silvered dial
Gold applied numerals
Movement triple calendar with moon phase Cal. 381

14 kt gold filled wristwatch "Cosmic"
Three-body case ref. DW2486-2, ø 37 mm.
Black dial for Iranian market
Gilt persian numerals
Movement triple calendar with moon phase Cal. 381

Royal Army Force stainless steel wristwatch
Screwed back case, ø 34.5 mm.
Black dial with broad arrow
Radium arabic numerals and dots
Cal. 30 T2

Silver chronograph pocket watch
Four body hinged case, ø 58 mm.
White enamel dial with five colours scales with brevet
A.T. tachimeter scale with black Arabic numerals
Cal. 19''' Chro

"Double face" stainless steel chronograph pocket watch
Four body hinged case, ø 58 mm.
White enamel dial with five colours scales with brevet.
A.T. tachymeter scale with black Arabic numerals
Cal. 19''' Chro

Early silver chronograph wristwatch
Three-body hinged case ø 44 mm.
Enamel dial
Black Arabic numerals
Cal. 18''' EP Chro

Early 18 kt gold chronograph wristwatch
Three-body hinged case, ø 44 mm.
Enamel dial
Black Arabic numerals
Cal. 18''' EP Chro

Early 18 kt gold chronograph wristwatch
Four-body hinged case, ø 44 mm.
Enamel dial with three colours scales
Tachymeter scale with black Arabic numerals
Cal. 18''' EP Chro

Early stainless steel chronograph wristwatch
Three-body case with hinged lugs, ø 42 mm.
Enamel dial
Black Arabic numerals
Mov. with two double buttons Cal. 130 39 Chro

Stainless steel chronograph wristwatch
Three-body case, ø 37 mm.
Enamel dial with four colours scales
Tachymeter scale with Arabic numerals
Cal. 33.3 Chro T1

Stainless steel chronograph wristwatch
Two-body case, ø 38 mm.
Enamel dial with four colours scales
Tachymeter scale with Arabic numerals
Cal. 33.3 Chro

Stainless steel chronograph wristwatch
Tonneau shaped three-body case, ø 33 mm.
Enamel dial with two colours scales
Tachymeter scale with Arabic numerals
Cal. 28.9 Chro T1

18 kt gold chronograph wristwatch
Tonneau shaped three-body case, ø 33 mm.
Enamel dial with four colours scales
Tachymeter scale with black Arabic numerals
Cal. 28.9 Chro T1

Stainless steel chronograph wristwatch
Three-body case, ø 37 mm.
Black dial
Tachymeter and telemeter gilt scales
Cal. 33.3 Chro T1

14 kt gold chronograph wristwatch
Three body case with engraving bezel, ø 38 mm.
Black dial
Tachymeter and telemeter gilt scales
Cal. 33.3 Chro T1

Stainless steel chronograph wristwatch
Screwed back case, ø 37 mm.
Two-tone silvered dial
Tachymeter, telemeter and pulsometer scale
Cal. 33.3 Chro T2

Stainless steel chronograph wristwatch
Screwed back case, ø 37 mm.
Black dial
Tachymeter, telemeter and pulsometer gilt scale
Cal. 33.3 Chro T2

Stainless steel chronograph wristwatch
Screwed back case ref CK2077-2, ø 38 mm.
Black dial
Radium Arabic numerals
Cal. 33.3 Chro T2

Stainless steel chronograph wristwatch
Screwed back case ref CK2077, ø 38 mm.
Grey slate dial mod. 2039
Tachymeter scales with Arabic numerals
Cal. 33.3 Chro T2

Stainless steel chronograph wristwatch
Three-body case ref. CK2393, ø 37 mm.
Black dial
Tachymeter and telemeter gilt scales
Cal. 33.3 Chro T1

Stainless steel chronograph wristwatch
Screwed back case ref CK2077, ø 38 mm.
Silvered dial
Tachymeter scales with gold Arabic and baton numerals
Cal. 33.3 Chro T2

14 kt gold chronograph wristwatch
Three-body case ref. OJ2393, ø 37 mm.
Black dial
Tachymeter and telemeter gilt scales
Cal. 33.3 Chro T1

Stainless steel chronograph wristwatch
Three-body case ref. CK2393, ø 37.7 mm.
Black dial
Tachymeter and telemeter scales with Roman numerals
Cal. 33.3 Chro T1

Stainless steel chronograph wristwatch
Three-body case ref. CK2393, ø 37.7 mm.
Silvered dial with four colours scales
Tachymeter, telemeter and pulsometer scales
Cal. 33.3 Chro T1

Stainless steel chronograph wristwatch
Three-body case ref. CK2393, ø 37.7 mm.
Silvered dial
Tachymeter scales with relief Arabic numerals,
Cal. 33.3 Chro T1

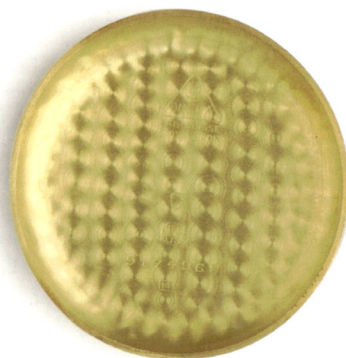

18 kt gold chronograph wristwatch
Three-body case ref. OT2393, ø 37.7 mm.
Silvered dial three colours scales
Tachymeter, telemeter and pulsometer scales
Cal. 33.3 Chro T1

Stainless steel chronograph wristwatch
Screwed back case ref. CK2451-7, ø 35 mm.
Silvered dial
Radium Arabic numerals and dots.
Cal. 321

Stainless steel chronometer wristwatch
Screwed back case ref. CK2254, ø 35 mm.
Silvered dial mod. 925
Arabic numerals
Cal. 30 SC T2

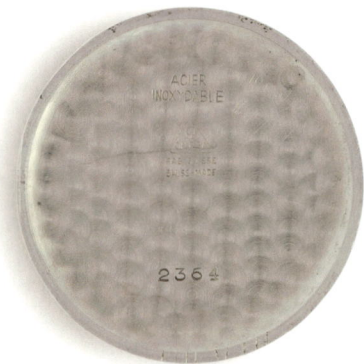

Stainless steel chronometer wristwatch
Three-body case ref. CK2364, ø 33 mm.
Two-tone silvered dial
Applied Roman numerals
Cal. 30 T2 RG

Stainless steel chronometer wristwatch
Three-body case ref. CK2364, ø 33 mm.
Two-tone silvered dial mod. 4038
Arabic numerals
Cal. 30 T2 RG

18 kt pink gold chronometer wristwatch
Three-body case ref. OT2367, ø 35.5 mm.
Ping gold dial
Arabic numerals
Cal. 30 SC T2 RG

18 kt pink gold chronometer wristwatch
Three-body case ref. OT2367, ø 35.5 mm.
Two-tone silvered dial ref. 7067
Gold applied Roman and baton numerals
Cal. 30 SC T2 RG

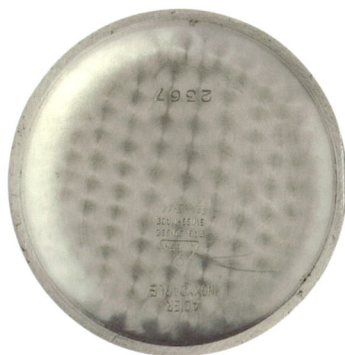

Stainless chronometer wristwatch
Three-body case ref. CK2367, ø 35.5 mm.
Two-tone silvered dial
Black Arabic numerals
Cal. 30 SC T2 RG

18 kt pink gold chronometer wristwatch
Three-body case ref. OT2364, ø 33 mm.
Two-toneSilvered dial
Gold applied Roman and baton numerals
Cal. 30 T2 RG

18 kt pink gold chronometer wristwatch
Three-body case ref. OT14169, ø 37 mm.
Two-tone silvered dial
Applied Roman and baton numerals
Cal. 30 T2 RG

Stainless steel and 14 kt gold chronometer wristwatch
Three-body case ref. CO2367, ø 35.5 mm.
Two-tone silvered dial
Gold applied Roman and baton numerals
Cal. 30 SC T2 RG

18 kt gold chronometer wristwatch
Three-body case ref. OT2365, ø 33 mm.
Silvered dial
Arabic numerals
Cal. 30 SC T2 RG

Stainless steel and 14 kt gold chronometer wristwatch
Three-body case ref. CO2366, ø 35.5 mm.
Silvered dial
Gold applied Arabic and baton numerals
Cal. 30 T2 RG

Desk clock chronometer gold plated brass and wood
Two-body case with transparent back ref. 5005/1, 145x145 mm.
Silvered dial with exterior second track
Applied Roman and baton numerals
Mov. 8 days true-beat seconds cal. 59 8-D SC ST1

Desk clock "Marine Megaquartz"
Three-body aluminium case
ref. 5802, Black dial
White numerals
Cal. 1525 "VEGA"

Platinum chronometer wristwatch "Co-Axial"
Screwed back ref. BT5921.61.31, ø 38 mm.
Pink dial
Applied obus numerals
Mov. with coaxial escapement Cal. 2500

Stainless steel split seconds chronograph pocket watch "Black Space"
Two-body case with screwed back and special black finishing, ø 64mm.
Anti-shock plastic case, dual level black dial with second exterior track
White Arabic numerals, bezel divided for 1/10 second
Cal. 205

Stainless steel chronograph
Three-body case with special black finishing, ref. MG6301
Red plastic support
Black dial with bezel divided for 1/5 second
Cal. 8208 A

Stainless steel wristwatch "Chronoquartz"
Two-body case with integrated bracelet ref. ST196.0052
Commemorative model for Montreal Olympic Games 1976
Analogic and digital display
Cal. 1611-Albatros-

Stainless steel chronostop wristwatch
Screw back case ref. ST146-10, ø 34 mm.
Dark red dial
Applied stainless steel indexes
Cal. 865

Gold filled 18 kt. self-winding wristwatch
Three-body case, ref. DW2713-2, ø 38 mm.
Argenté guillochè dial
Applied gold numerals
Cal. 30 T2 PC

Stainless steel self-winding wristwatch
Screw back case ref. CK2494, ø 36 mm.
Argenté dial with grey dark external track
Radium roman numerals
Cal. 330

18 kt. pink gold self-winding wristwatch
Three-body case ref. OT2714, ø 37 mm.
Argenté dial
Applied gold Arabic and baton numerals
Cal. 344

Gold filled 10 kt. self-winding wristwatch
Screw back case, ref. GX6279, ø 35 mm.
Argenté dial with AMF logo
Applied gold Arabic and baton numerals
Cal. 500

18 kt. pink gold self-winding wristwatch
Three-body case ref. OT2546, ø 37 mm.
Cloisonné enamel dial
Applied gold baton numerals
Cal. 333

18 kt. pink gold self-winding wristwatch
Three-body case ref. OT2879, ø 37 mm.
Guillochè dial
Applied gold baton numerals
Cal. 491

Stainless steel self-winding wristwatch
Two-body case ref. CK3903-2, 45x33 mm.
Argenté dial
Applied stainless steel baton numerals
Cal. 342 SC

Stainless steel self-winding wristwatch
Two-body case ref. CK3903-2, 45x33 mm.
Argenté dial
Sector dial
Cal. 342 SC

Platin self-winding wristwatch
Two-body case ref. PA14.607, 42x33 mm.
Present from Caudillo Francisco Franco Bahamonde
to Mexican President Miguel Alemàn Valdes
Guilloché dial with diamonds, Cal. 342

Stainless steel self-winding wristwatch
Two-body case, brac.Omega, ref. CK3903-2, 45x33 mm.
Argenté dial
Applied stainless steel Arabic numerals
Cal. 342

18 kt. gold self-winding wristwatch
Two-body case, brac.Omega, ref. OT14607, 40x33 mm.
Argenté dial
Applied gold numerals
Cal. 342

18 kt. gold self-winding wristwatch
Two-body case, 31x31 mm.
Cloisonné enamel dial (American Flag)
Black baton hands
Cal. 471

18 kt. gold self-winding wristwatch
Three-body case, brac. Omega ref. OT2848, ø 34 mm.
Globemaster (first American market name for the Constellation model)
Black dial with radium Arabic numerals
Cal. 501

Stainless steel self-winding wristwatch "Genève"
Three-body case ref. CK2982-5, ø 34 mm.
Two-tone argenté dial
Applied gold baton numerals
Cal. 503

Stainless steel wristwatch "Seamaster"
Three-body case ref. CK2767-10, ø 34 mm.
Argenté dial
Applied numerals
Cal. 354

Stainless steel wristwatch "Seamaster"
Three-body case, brac. Omega, ref. CK 2576-2, ø 34 mm.
Argenté dial
Applied Breguet numerals
Cal. 344

Pink gold 18 kt wristwatch "Seamaster"
Three-body case, brac. Omega, ref. OT2848, ø 34 mm.
Argenté dial
Applied gold baton numerals
Cal. 501

Stainless steel wristwatch "Seamaster"
Three-body case with hooded lugs, brac. Omega, ref. CK14363-5, ø 34 mm.
Argenté dial
Applied Arabic and baton numerals
Cal. 501

Stainless steel wristwatch "Seamaster"
Three-body case, brac. Omega, ref. CK2976-2, ø 38 mm.
Two tone argenté dial
Applied stainless steel baton numerals
Cal. 501

18 kt gold wristwatch "Seamaster"
Three-body case, brac. Omega, ref. OT2848, ø 34 mm.
Argenté dial
Applied gold baton numerals
Cal. 501

18 kt gold wristwatch "Seamaster"
Three-body case, brac. Omega, ref. OT2848, ø 34 mm.
Black dial
Applied gold Arabic and baton numerals
Cal. 501

Stainless steel wristwatch "Seamaster"
Three-body case, ref. CK 2577-12, ø 34 mm.
Guilloché dial
Applied gold Arabic and baton numerals
Cal. 352

18 kt gold wristwatch "Seamaster"
Three-body case, ref. OT2520, ø 34 mm.
Cloisonné enamel dial mod. Neptune chariot
Gold applied baton numerals
Cal. 354

Platin wristwatch "Seamaster"
Three-body case with hooded lugs, ref. PA14326, ø 36 mm.
Guilloché dial
Applied diamonds numerals
Cal. 352

Stainless steel wristwatch "Seamaster 300"
Screwed back case with revolving bezel ref. ST14755-62, ø 39 mm.
Black dial, mod. 8238, "Broad arrow" hand
Luminous baton numerals
Cal. 552

Stainless steel wristwatch "Railmaster"
Screwed back case, ref. CK2914-4, ø 38 mm.
Black dial, mod. 8238, "Broad arrow" hand
Luminous baton numerals
Cal. 284

Stainless steel wristwatch "Railmaster"
Screwed back case, ref. CK2914-6, ø 38 mm.
Black dial, mod. 8238, "Broad arrow" hand
Luminous baton numerals
Cal. 284

Stainless steel wristwatch "Seamaster 300"
Screwed back case with revolving bezel ref. ST166.024, ø 40
Black dial
Luminous baton numerals
Cal. 565

147

Stainless steel wristwatch "Seamaster 300"
Screwed back case with revolving bezel ref. ST165.024, ø 40 mm.
Blue dial
Luminous baton numerals
Cal. 552

Stainless steel chronograph wristwatch "Seamaster"
Two-body case, ref. ST145.016, ø 38 mm.
Black dial
Tachimeter scale with applied baton numerals
Cal. 861

Stainless steel wristwatch "Seamaster 600"
Waterproof case with revolving bezel ref. ST166.077, ø 44 mm.
Blue dial
Luminous baton numerals
Cal. 1002

Titanium wristwatch "Seamaster 600"
Revolving bezel ref. ST166.077, ø 44 mm.
Prototype manufactured in 12 pieces
Blue dial with luminous baton numerals
Cal. 1002

Stainless steel wristwatch "Seamaster 600"
Waterproof case with revolving bezel, brac. Omega with
matt grey special finishing ref. ST166.077, ø 44 mm.
Blue dial with luminous baton numerals
Cal. 1002

Stainless steel wristwatch "Seamaster 1000"
Waterproof case with revolving bezel, ref. ST166.0093, ø 42 mm.
Brac. Omega mod. ST1266/237
Blue dial with luminous baton numerals
Cal. 1002

Stainless steel wristwatch "Seamaster 200"
Waterproof case with revolving bezel, ref. ST166-091, ø 41 mm.
Black dial
Luminous baton numerals
Cal. 1012

Stainless steel wristwatch "Seamaster"
Waterproof case with revolving bezel, brac. Omega ref. ST3660858, ø 41 mm.
Blue dial
Luminous baton numerals
Cal. 1012

Stainless steel chronograph wristwatch "Seamaster"
Two-body black chamfered carbide case ref. 145.0023, ø 44 mm.
Brac. Omega mod. ST1266/237, dark grey dial with two colours counters
Tachimeter scale with luminous baton numerals
Cal. 861

Stainless steel chronograph wristwatch "Seamaster"
Two-body chamfered carbide case ref. 145.0023, ø 44 mm.
Brac. Omega mod. ST1266/237, black dial with two colours counters
Tachimeter scale with luminous baton numerals
Cal. 861

18 kt. gold chronograph wristwatch
Two-body waterproof case ref. OT14364, ø 36 mm.
Argenté dial, tachimeter scale
Applied gold baton numerals
Cal. 321

Stainless steel chronograph wristwatch
Three-body case, ø 36 mm.
Argenté dial, tachymeter scale
Applied baton numerals
Cal. 321

Stainless steel chronostop wristwatch
Two-body case ref. 146.10, ø 41 mm.
Black dial with tachymeter scale
Applied baton numerals
Cal. 865

Stainless steel chronograph wristwatch
Waterproof case with revolving bezel, ref. ST176.004, ø 43 mm.
Brac. Omega mod. ST1266/237
Blue dial with luminous baton numerals
Cal. 1040

Stainless steel wristwatch "Seamaster 200"
Waterproof case with revolving bezel, ref. ST166.068, ø 41 mm.
Dark grey dial
Luminous baton numerals
Cal. 565

Stainless steel wristwatch "Memomatic"
Two-body waterproof case, ø 41 mm.
Grey dial
Luminous numerals
Mov. with alarm cal. 980

Stainless steel wristwatch "Seamaster"
Two-body case with integrated bracelet "Armadillo" ref. ST198.042
Brevet 508925 brac. mod. 112/203, ø 38 mm.
Argenté dial, applied baton numerals
Cal. 1260 "SOURIS" Megasonic F300Hz

Stainless steel wristwatch "Speedsonic"
Two-body case with integrated bracelet "Armadillo" ref. ST388.0001
Brac. mod. 112/203, ø 44 mm.
Argenté dial with tachimeter scale, applied baton numerals
Cal. 1255

Stainless steel chronograph wristwatch
Waterproof case with internal revolving bezel" ref. ST146-011, 50x16 mm.
Grey dial with three colours minutes counter
Applied numerals
Cal. 930

Stainless steel wristwatch
"Seamaster Professional 300" Waterproof case
with revolving bezel, ref. ST2531-80, ø 40 mm.
Blue guilloché dial, luminous baton numerals
Cal. 1120

Stainless steel wristwatch "Constellation"
Three-body case, brac. Omega ref. CK2852-8, ø 34 mm.
Two-tone argenté dial "pie pan"
Applied stainless steel arabic numerals and dot markers
Cal. 505

18 kt gold wristwatch "Constellation"
Three-body case, brac. Omega ref. OT2852, ø 34 mm.
Two-tone argenté dial
Applied gold Arabic numerals and dot markers
Cal. 505

18 kt gold wristwatch "Constellation"
Three-body case, brac. Omega ref. OT2652, ø 34 mm.
Black dial "pie pan"
Applied gold dot markers
Cal. 354

Platinum wristwatch "constellation de Luxe"
Three-body case, ref. PA2852, ø 34 mm.
Two-tone argenté dial "pie pan"
Applied platinum dot markers
Cal. 505

14 kt gold wristwatch "Constellation"
Three-body case, ref. OJ2852, brac. Omega ref. 8210, ø 34 mm.
Guilloché black dial "pie pan"
Applied gold dot markers
Cal. 354

18 kt gold wristwatch "Constellation de Luxe"
Three-body case, ref. OT2852 brac. Omega mod. 6628/18, ø 34 mm.
Two-tone gold dial "pie pan"
Applied gold dot markers
Cal. 505

181

18 kt pink gold wristwatch "Constellation"
Three-body case, ref. OT2852, brac. Omega, ø 34 mm.
Two-tone argenté dial
Applied gold arabic numerals and dot markers
Cal. 505

18 kt pink gold wristwatch "Constellation de Luxe"
Three-body case, ref. OT2852, brac. Omega Mod. 6622/18, ø 34 mm.
Two-tone gold dial "pie pan"
Applied gold dot markers
Cal. 505

18 kt white gold wristwatch "Constellation de Luxe"
Three-body case, ref. OG2852, brac. Omega, ø 34 mm.
Two-tone argenté dial "pie pan"
Applied white gold dot markers
Cal. 505

18 kt yellow gold wristwatch "Constellation de Luxe"
Three-body case, ref. OT2852, brac. Omega mod. 6622/18, ø 34 mm.
Two-tone gold dial "pie pan"
Applied white gold dot markers
Cal. 505

18 kt yellow gold wristwatch "Constellation Gran de Luxe"
Three-body case, ref. OT2852, brac. Omega (snake), ø 34 mm.
Two-tone gold dial "pie pan"
Applied white gold dot markers
Cal. 505

18 kt pink gold wristwatch "Constellation de Luxe"
Three-body case, ref. OT2943, ø 34 mm.
Two-tone gold dial "pie pan"
Applied white gold dot markers
Cal. 505

18 kt pink gold wristwatch "Constellation de Luxe"
Three-body case, ref. OT14394, ø 34 mm.
Gold dial mod. 8013
Applied gold baton numerals
Cal. 501

Stainless steel wristwatch "Constellation"
Three-body case, brac. Omega ref. ST14394, ø 34 mm
Striped dial
Applied stainless steel baton numerals
Cal. 501

18 kt gold wristwatch "Constellation Gran de Luxe"
Screwed back case with hooded lugs, brac. Omega (snake) ref. OT14355, ø 35 mm.
Two-tone gold dial "pie pan"
Applied gold dot markers
Cal. 354

18 kt gold wristwatch "Constellation de Luxe"
Three-body case with hooded lugs ref. OT3930, ø 35 mm.
Two-tone gold dial "pie pan"
Applied gold dot markers
Cal. 505

18 kt gold wristwatch "Constellation Gran de Luxe"
Screwed back case with hooded lugs, brac. Omega (snake) ref. OT14355, ø 35 mm.
Two-tone gold dial "pie pan"
Applied gold baton numerals and diamonds
Cal. 505

18 kt white gold wristwatch "Constellation Gran de Luxe"
Screwed back case with hooded lugs, brac. Omega (snake) ref. OG14355, ø 35 mm.
Two-tone gold dial "pie pan"
Applied gold dot markers
Cal. 354

18 kt gold wristwatch
Three-body case, brac. Omega ref. OT2842, ø 34 mm
Cloisonné enamel dial mod. Fantasy
Applied gold baton numerals
Cal. 330

18 kt gold wristwatch "Constellation"
Three-body case, brac. Omega ref. OT2652, ø 34 mm
Cloisonné enamel dial mod. Observatory
Applied gold baton numerals
Cal. 330

18 kt pink gold wristwatch "Constellation"
Three-body case, ref. OT2782, ø 34 mm
Cloisonné enamel dial mod. Observatory
Applied gold baton numerals
Cal. 354

18 kt gold wristwatch "Constellation"
Three-body case, brac. Omega ref. OT2782, ø 34 mm
Cloisonné enamel dial mod. Observatory
Applied gold baton numerals
Cal. 330

18 kt. gold wristwatch "Constellation mod. C"
Screwed back case, brac. Omega ref. BA168011, 35x41 mm.
Gilt dial
Applied gold baton numerals
Cal. 565

18 kt. white gold wristwatch "Constellation mod. C"
Screwed back case, brac. Omega ref. BC168011, 35x41 mm.
Argenté dial
Applied diamond baton numerals
Cal. 564

18 kt gold wristwatch "Constellation"
Two-body case, closed by 4 screws, ref. BA191.0003, 44x34 mm.
Present from Omega to Ursula Andress, 1973
Gilt dial, amber quartz glass, painted baton numerals
Cal. 1302-BETA 21-Elettroquartz F 8192Hz

18 kt. white gold wristwatch "Constellation mod. C"
Screwed back case, brac. Omega ref. BC168011, 35x41 mm.
Argenté dial
Applied diamond baton numerals
Cal. 564

18 kt. white gold wristwatch "Constellation mod. C"
Screwed back case, brac. Omega ref. BC168029, 35x41 mm.
Argenté dial
Applied diamond baton numerals
Cal. 751

18 kt. white gold wristwatch "Constellation mod. C"
Screwed back case, brac. Omega ref. BC168029, 35x41 mm.
Grey dial
Applied luminous baton numerals
Cal. 751

18 kt. white gold wristwatch "Constellation mod. C"
Screwed back case, brac. Omega ref. BC168017, 35x41 mm.
Argenté dial
Applied luminous baton numerals
Cal. 564

18 kt. white gold wristwatch "Constellation"
Screwed back case, brac. Omega ref. BC168004-4, ø 35 mm.
Argenté dial
Applied white gold and diamond baton numerals
Cal. 561

18 kt. white gold wristwatch "Constellation"
Two-body case, closed by 4 screws, brac. Omega ref. BC8359, 48x30 mm.
Grey dial
Applied white gold baton numerals
Cal. 711

18 kt. white gold wristwatch "Constellation"
Two-body case, closed by 4 screws, brac. Omega, ref. BC8539, 48x30 mm.
Grey dial
Applied white gold baton numerals
Cal. 711

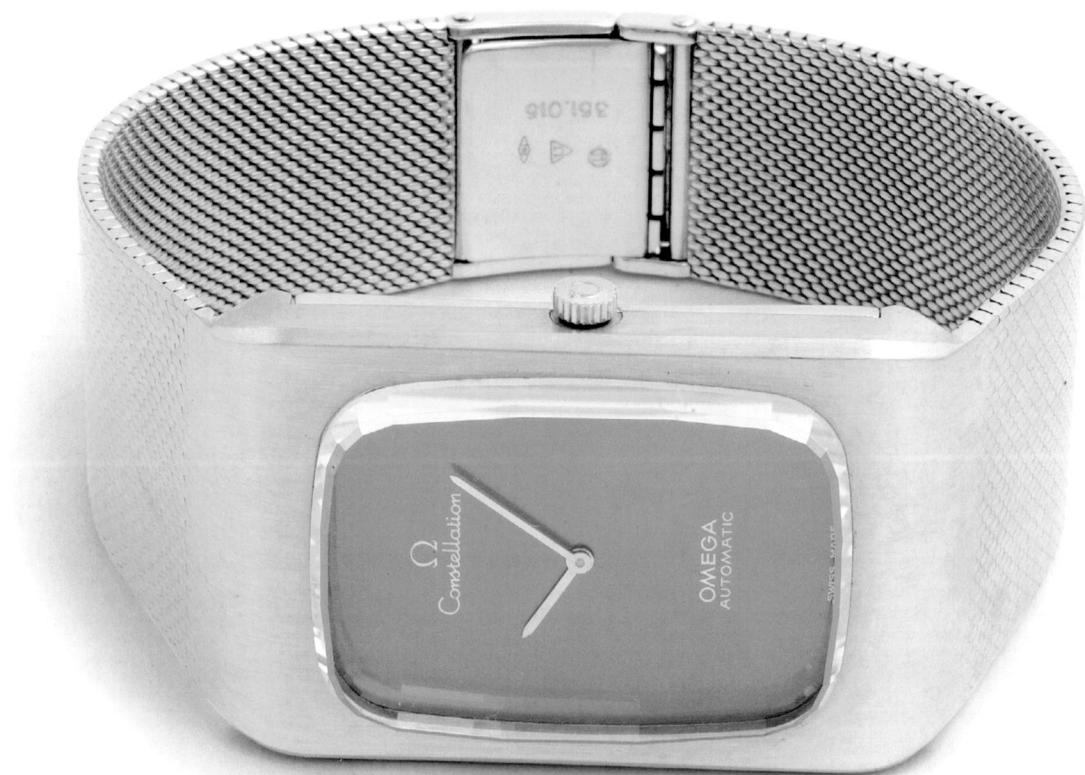

18 kt. white gold wristwatch "Constellation"
Two-body case brac. Omega ref. BC8354, 42x30 mm.
Two-tone grey dial
White gold hands
Cal. 711

18 kt. white gold wristwatch "Constellation mod. C"
Two-body case, closed by 4 screws, ref. BC191.0003, 44x34 mm.
Argenté dial
Painted baton numerals
Cal. 1302- BETA 21-Elettroquartz F 8192Hz

18 kt gold wristwatch "Constellation"
Two-body case, closed by 4 screws, ref. BA191.0003, 44x34 mm.
Present from Omega to Ursula Andress, 1973
Gilt dial, amber quartz glass, painted baton numerals
Cal. 1302-BETA 21-Elettroquartz F 8192Hz

18 kt. white gold wristwatch "Constellation"
Two-body case, closed by 4 screws, ref. BC191.0001, 44x36 mm.
Argenté dial
Painted baton numerals
Cal. 1302-BETA 21-Elettroquartz F 8192Hz

18 kt. white gold wristwatch "Constellation"
Two-body case, closed by 4 screws, ref. BA191.0001, 48x30 mm.
Gilt dial
Amber quartz glass
Cal. 1302- BETA 21-Elettroquartz F 8192Hz

18 kt. white gold wristwatch "Constellation"
Two-body case, closed by 4 screws, ref. BA191.0003, 44x34 mm.
Gilt dial
Painted baton numerals
Cal. 1302- BETA 21-Elettroquartz F 8192Hz

18 kt. white gold wristwatch "Constellation"
Two-body case, closed by 4 screws, brac. Omega, 44x30 mm.
Gilt dial
Amber quartz glass
Cal. 1302-BETA 21-Elettroquartz F 8192Hz

18 kt. white gold wristwatch "Constellation"
Two-body case, closed by 4 screws, brac. Omega, 51x30 mm.
Argenté dial
Painted baton numerals
Cal. 1302-BETA 21-Elettroquartz F 8192Hz

OBSERVATOIRE

SERVICE CHRONOMETRIQUE

Bulletin de Marche n° *866*

délivré le *9 Novembre 1976*

au CHRONOMÈTRE DE MARINE

Fabricant .... OMEGA

Calibre 1516

Mouvement n° *37 058 381*

Particularités .... Quartz 2,4 MHz

Vu et approuvé, Le Chef du Service :

Tableau des principaux critères	...vées	Valeurs limites
marche diurne maximale pa...	*81*	± 3,000 s/d
écart moyen maximal	*0,011*	0,500 s/d
écart moyen de la m...	*0,003*	0,170 s/d
écart moyen de ...	*0,010*	1,000 s/d
erreur primaire...	*+ 0,013*	± 0,070 s/d (°C)⁻¹
erreur sec...	*− 0,090*	± 1,000 s/d
erreur ...	*+ 0,005*	± 1,500 s/d

— : retard   s/d : seconde par jour

...B. A la diligence du Fabricant, un extrait du registre des comparaisons de l'Observatoire peut être joint à ce bulletin de marche.

NEO-TYPO - BESANÇON

Stainless steel and 14 kt gold wristwatch "Marine Chronometer"
Integrated bracelet and case, ref. 398 0836, 48x30 mm.
Black dial
Luminous baton numerlas
Cal. 1516-Megaquartz F 2.4 Mhz

18 kt. gold wristwatch "Constellation"
Two-body case, ref. BA396.0802, 42x36 mm.
Gilt dial"
Painted baton numerals
Cal. 1303-BETA 21-Elettroquartz F 8192Hz

Stainless steel chronograph wristwatch, First model of Speedmaster
Screwed back case, stainless steel bezel, ref. CK2915.1, ø 39 mm.
Brac. Omega mod. 7077, black dial
Radium baton numerals and "Broad arrow" hand
Cal. 321

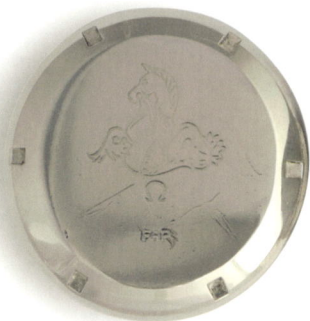

Stainless steel chronograph wristwatch "Speedmaster"
Screwed back case and stainless steel bezel, ref. CK2915.2, ø 39 mm.
Model for "FAP" Fuerza Aerea of Perù, brac. Omega Mod. 7077,
Black dial, radium baton numerals and "Broad arrow" hand
Cal. 321

Stainless steel chronograph wristwatch "Speedmaster"
Screwed back case and stainless steeel bezel, ref. CK2915.2, ø 39 mm.
Brac. Omega Mod. 7077
Black dial, radium baton numerals and "Broad arrow" hand
Cal. 321

Stainless steel chronograph wristwatch "Speedmaster"
Screwed back case ref. CK2998.1, ø 39 mm.
Brac. Omega mod. 7077, Bezel with tachymeter scale base 1000
Black dial, radium baton numerals and "Broad arrow" hand
Cal. 321

Stainless steel chronograph wristwatch "Speedmaster"
Screwed back case ref. CK2998.5, brac. Omega mod. 7912, ø 39 mm.
Bezel with tachimeter scale base 500
Black dial, radium baton numerals and "alpha" hands
Cal. 321

Stainless steel chronograph wristwatch "Speedmaster"
Screwed back case ref. ST105.002, ø 39 mm.
Bezel with tachymeter scale base 500
Black dial, radium baton numerals and "alpha" hands
Cal. 321

Stainless steel chronograph wristwatch "Speedmaster"
Screwed back case ref. ST105.003, brac. Omega mod. 1501, ø 40 mm.
Bezel with tachymeter scale base 500
Black dial, radium baton numerals and "baton" hands
Cal. 321

Stainless steel chronograph wristwatch "Speedmaster Professional"
Screwed back case ref. ST105.012, ø 42 mm.
Bezel with tachymeter scale base 500
Black dial, luminous baton numerals and "baton" hands
Cal. 321

Stainless steel chronograph wristwatch "Speedmaster Professional"
Screwed back case ref. ST145.0022, ø 42 mm.
Bezel with pulsometer scale base 15
Black dial, luminous baton numerals and stick "baton" hands
Cal. 861

Stainless steel chronograph wristwatch "Speedmaster Professional"
Screwed back case ref. ST145.0022, ø 42 mm.
Commemorative model in 500 pieces, for the test Apollo-Soyuz 1975
Black dial, luminous baton numerals and "baton" hands
Cal. 861

Stainless steel chornograph wirstwatch "Speedmaster Professional"
Screwed back case ref. ST145.012, ø 42 mm
Bezel with tachymeter scale base 500
Blue dial, luminous baton numerals and "baton" hands
Cal. 321

Stainless steel chronograph wristwatch "Speedmaster Professional Mark II"
Screwed back case ref. ST145.014, brac. Omega mod. ST1116/173, ø 42 mm.
Tachymeter scale base 500
Black dial, luminous baton numerals and "baton" hands
Cal. 861

Stainless steel chronograph wristwatch "Speedmaster Professional Mark III"
Screwed back case, ref. ST176.0002, brac. Omega mod. 1162/172, ø 41 mm.
Bezel with tachymeter scale base 500
Blue dial with applied baton numerals
Cal. 1040

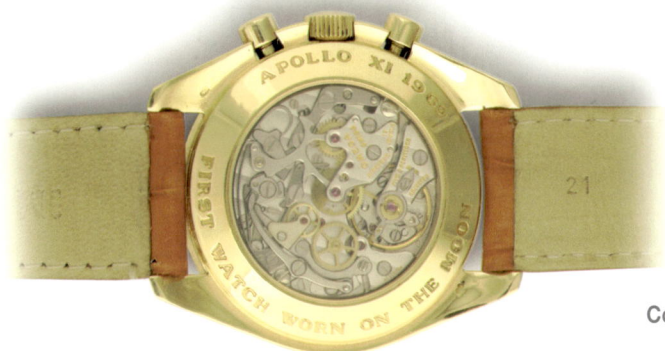

18 kt gold chronograph wristwatch "Speedmaster Professional"
Screw back case, ref. BA145.0039, ø 42 mm.
Commemorative model, for the 10th of the landing on the moon 1980
Champagne dial with black baton numerals
Cal. 861L rodium movement

Stainless steel chronograph wristwatch "Flightmaster"
First model, screwed back case, brac. Omega ref. ST145.013, ø 42 mm.
Black dial, counter 24 h, revolving bezel 60 mn.
Luminous baton numerals
Cal. 910

Stainless steel chronograph wristwatch "Flightmaster"
Second model, screwed back case, brac. Omega ref. ST145.036, ø 42 mm.
Black dial, revolving bezel 60 mn.
Luminous baton numerals
Cal. 911

Titanium chronograph wristwatch "Speedmaster"
Two-body case with brac. Omega, ref. TI1345.0810, ø 43 mm.
Limited edition 400 pieces, 1986
Black dial, applied baton numerals
Mov. with date and moonphases Cal. 866

Stainless steel chronograph wristwatch "Speedmaster"
Screw back case, ref. ST145.0809, ø 42 mm.
Limited edition 1300 pieces, 1985
Black dial, luminous baton numerals
Cal. 866

18 kt white gold chronograph wristwatch "Speedmaster Professional"
Screwed back case, brac. Omega ref. BC348.0062, ø 42 mm.
Commemorative model, manufactured in 500 pieces,
for the 25th anniversary of the landing on the moon 1994
Argenté dial with applied baton numerals
Cal. 864

Titanium multifunction wristwatch "X-33"
Three-body case, revolving bezel, ref. TS186.1998, ø 44 mm.
Luminescent analogic digital dial
Luminous baton numerals
Cal. 1666 Quartz

"Speedmaster Missions" case limited edition (40 pcs)
23 stainless steel wristwatches "Speedmaster"
ø 42 mm. plus a mov. Cal. 186 for the 40th Speedmaster
Anniversary. 22 wristwatches with Mission patch
on dial from 1965 to 1973 Ref. ST345.0022,
plus one "replica" of the first speedmaster
1957 model (SS345.0022)

Gemini V 21-29 August 1965,
Gemini VI 15-16 December 1965,
Gemini VII 4-18 December 1965,
Gemini VIII 16 march 1966
Gemini IX 3-6 June 1966
Gemini X 18-21 July 1966
Gemini XI 12-15 September 1966
Gemini XII 11-15 November 1966

Apollo 7 11-22 November 1968
Apollo 8 21-27 December 1968
Apollo 9 3-13 March 1969
Apollo 10 18-26 May 1969
Apollo 11 16-24 July 1969
Apollo 12 14-24 November 1969
Apollo 13 11-17 April 1970
Apollo 14 31 January-9 February 1971

Apollo 15 26 July-7 August 1971
Apollo 16 16-27 Aprile 1972
Apollo 17 7-19 December 1972
Skylab 1 25 May-22 June 1973
Skylab 2 23 July-25 September 1973
Skylab 3 16 November 1973-8 February 1974

245

Stainless steel wristwatch "Speedmaster Professional"
Three-body case, brac. Omega ref. 178.0031, ø 42 mm.
Commemorative model, manufactured in 6000 pieces,
For the 6th title of Michael Schumacher F1 driver championship, 2004
Cal. 3301

Stainless steel wristwatch "Speedmaster Professional"
Three-body case, ref. ST145.0022, brac. Omega mod. 1998, ø 42 mm.
Limited edition, manufactured in 2004 pieces for the Japanese market, 2004
Grey dial with burgundy and orange indexes.
Cal. 1861

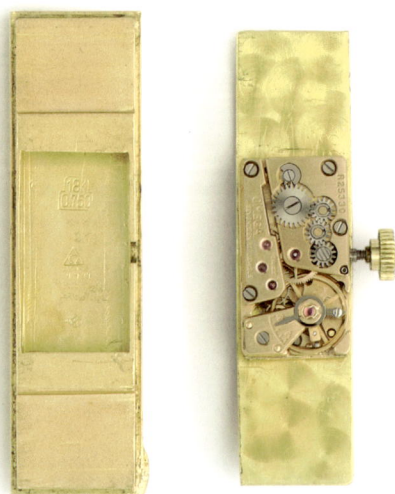

18 kt. gold wristwatch
two-body pieces, hinged lugs manufactured for Boucheron, 58x16 mm.
Gilt dial, 18 kt gold closing clasp
Black baton numerals
Cal. 730

18 kt. gold wristwatch
Two-body case oval shaped, closed by 4 screws, 31x20 mm.
Argenté dial
Black roman numerals
Cal. 620

18 kt. gold wristwatch "Dynamic"
Two-body case, brevet nr. 529.376, 35x38 mm.
Bracelet made with Corfam brevet nr. 569.319
Two-tone gilt dial with luminous baton numerals
Cal. 565

Stainless steel wristwatch "Dynamic"
Two-body case, brevet nr. 529.376, 35x38 mm.
Bracelet made with natural leather
White dial with luminous baton numerals
Cal. 565

Ceramic wristwatch "Art"
Two-body case closed by four screws, ref. 196.0.440, ø 39 mm.
The back is decorated with a painting of Al Held
Limited edition 900 pieces, 1986
Black dial with date, Cal. ETA 255.141

Ceramic wristwatch "Art"
Two-body case closed by four screws ref. 196.0.440, ø 39 mm.
The back is decorated with a painting of Max Bill
Limited edition 900 pieces, 1986
White dial with date, Cal. ETA 255.141

18 kt white gold wristwatch "Art"
Two-body case closed by four screws, ref. BC196.0.440, ø 39 mm.
The back is decorated with a painting of Paul Talman
Limited edition 900 pieces, 1986
Black dial, Cal. ETA 255.431